手绘上海经典建筑

张安朴 著

上海人民美术出版社

百年上海的建筑拼图

老友张安朴是位多才多艺、成果丰硕的画家。他的宣传画作品在第一届和第二届全国宣传画展览中均获得一等奖，他为上海电视节、上海国际电影节连续设计的海报成为当时上海影视节活动的形象符号，他独具个性的钢笔水彩画在上海水彩画界更是无人不晓。安朴近日编绘了一本钢笔水彩画册《手绘上海经典建筑》，即将付梓出版之际，他嘱我们兄弟俩为画册写一篇短文。

艺术与建筑，是我们与安朴的共同爱好。他经常会聊起我俩参与的一本介绍上海徐汇区老房子的书——《梧桐树后的老房子》，言语间流露出对上海老建筑的谙练。在我俩策划组织的"徐汇区老房子水彩画展"和"衡复老房子写生展"中，安朴都是位不可或缺的热情参与者。每次观摩上海老房子时，他时不时会画上几笔，并用文字记下要点，这是他在写生中养成的一个习惯。

有人说，百年历史看上海，这是因为上海的老建筑见证了中国近代史发展的重要时刻。20世纪初，西方各国看重上海的地理位置，预见这座港口城市将会发展成东方的巴黎，纷纷斥巨资在上海发展房产，很快在外滩一带及浦西建成颇具规模的建筑群。建筑作为城市中的容器，不仅是人们居住栖息之地，还承载着城市的历史与文化。工业崛起、经贸往来、政治变革、文化交流，在"容器"中或明或暗地完成，在冲突碰撞中达到融合。因而每一座建筑的背后都有着利益的博弈、情感的纠结，有着光鲜亮丽的一面，也有藏污纳垢的另一面。建筑作为空间中的实体，又显现出各具魅力的艺术风格：被称为"梧桐树后的老房子"的欧式洋房糅进了中国文化元素，成为上海地域建筑标志的石库门亦是中西合璧的。改革开放后，上海城市建设开启了脱胎换骨般的更新改造，而浦东的开发开放又催生了摩天大厦建筑群在黄浦江东岸拔地而起，与浦西外滩的"万国建筑博览会"相映成趣。对西方近代建筑的修缮保护，对本土优秀传统建筑的修复还原，赋予了这座繁华的国际大都市新生与活力。

张安朴的《手绘上海经典建筑》，以绘画的形式将上海百年建筑风采——绘出，附言简明扼要、点景入典。如同他的艺术设计作品，这本画册也是以线、点、面做结构布局。第一部分将静安寺、南京路商业街与外滩景观串联起来，形成一条经典的旅游线路。第二部分将散布在上海各处的30座邬达克设计的老建筑全盘托出，这30幅画作在"符号上海——当艺术遇见邬达克"展览结束后被悉数捐赠给了上海城市规划展示馆。第三部分以三个点分别描绘了上海最具代表性的古典园林建筑——豫园。他的作品《豫园》曾被选用于2013年中国邮政特种邮票。而嘉定与南翔古镇是画家"乡愁"的落点，他在画中倾泻出满满的情与爱。第四部分以面的形式回放了2010年上海世博会举办的盛况，展示了21世纪上海城市文化融汇天下的胸怀。阅读建筑，可以聆听从建筑的外表深入内中的人文故事，可以了解视觉与情感交融的过程，兴许还可以体会城市发展与历史保留难题的解析之术。所以说，《手绘上海经典建筑》不只是一本旅游和打卡景点的导览绘本，还具有传播城市文化的意义。

钢笔水彩画以线条造型为骨，以赋色为体，兼具速写和水彩画的特性，这是张安朴绘画艺术中的强项。他善于灵活运用硬笔与毛笔的不同性能来描绘建筑，看似随意的钢笔线条准确地勾画出建筑物的构造美感，且透视准确；饱含水分的色彩在冷暖对比中表现出建筑物的阴阳背向，还留下了好看的水渍。在《上海外滩》中，他隐去了钢笔线条，彰显水彩画的韵味；在《长桌街宴品小笼》中，他又以黑白对比的敷色手法，来衬托白描的街宴盛景。安朴在画中对作为点缀的人物基本不设色，着意于表现人物的生动姿态。在留白的运用上，他更是将这种虚实相济、以少胜多的艺术处理手段发挥得淋漓尽致，在上海水彩画界独树一帜。

张安朴用他的招牌式艺术风格记录下不同时代各具特色的建筑，还原出百年来上海城市的风姿神韵，让读者一册在手就能欣赏到绘画与建筑的美，不亦乐乎！

朱国荣（上海市美术家协会顾问、
上海市美术家协会原副主席兼秘书长）
朱志荣（徐汇区住房保障和房屋管理局原局长）
2021年9月

Sketch Book of Shanghai Buildings from the Past Century

Our old friend Zhang Anpu is a versatile and prolific painter. His posters won first prize in both thefirst and second national poster exhibitions and, for a number of years, he designed posters for the ShanghaiTV Festival and Shanghai International Film Festival during that time; his unique pen and ink with watercolor painting style is well known in Shanghai watercolor painting circles. When his pen and ink watercolor album Shanghai Sketchbook was about to bepublished, he asked us two brothers to write the preface for it.

Art and architecture are hobbies we have in common with Zhang Anpu. He often talks about ourbook Old Houses Behind Plane Trees in which we introduced the old historic houses in Xuhui District, Shanghai, which shows his deep knowledge of Shanghai's old buildings. In the "Watercolor Painting Exhibition of Xuhui District Historical Architecture" and "Painting Exhibition of the Old Buildings of Hengshan Road and Fuxing Road" that we planned and compiled, Zhang Anpu was an indispensable and enthusiastic participant. Every time he visited the old buildings of Shanghai, he would draw a few strokes and write down some key notes, a habit he developed in his sketching craft.

Some people say that if you want to see a hundred years history of china, you can simply look at Shanghai. This is because the old buildings in Shanghai have witnessed the great moments in the development of modern Chinese history. At the beginning of the 20th century, western countries valued the geographical location of Shanghai and predicted that the port city would develop into a Paris in the East. They spent a lot of money developing real estate in Shanghai and soon created a large-scale buildingcomplex on the Bund along the west side of the Huangpu River. As a container in the city, urban architecturerepresents not only a place where people work and reside, it also embodies and preserves the history andculture of the city. The rise of industry, economic and trade exchanges, political changes, and culturalexchanges are concluded, either explicitly or implicitly, within these "containers", and their integrationis achieved through conflict and collision. Therefore, behind each building, there is an evolving battle ofinterests and emotional entanglements and each building has both a bright side as well as another sidethat hides dirt. Chinese and Western mixed-style buildings, which are called "the old houses behind planetrees", are rife with Chinese cultural elements. Shikumen, which are the symbol of Shanghai-style regionalarchitecture, are also a combination of China and Western cultures. After the reform and opening-up, theurban construction of Shanghai was transformed, and the development and opening-up of Pudong then ledto the emergence of skyscraper buildings on the east bank of the Huangpu River, which contrast sharplyin style with the Shanghai Customs House building in Puxi on the west bank of the river. The repair andpreservation of modern western buildings and the restoration of local outstanding traditional buildings havegiven new life and vitality to this most prosperous international metropolis in the Far East.

Zhang Anpu's Shanghai Sketchbook depicts the century-old architectural style of Shanghai in theform of hundreds of individual paintings. The postscripts are brief and to the point, and the words thatZhang incorporated into the paintings perfectly compliment and enhance the visual aspects of the paintings.Zhu Guorong (Vice Chairman and Secretary General of Shanghai Artists Association)Zhu Zhirong (former Director of Xuhui District Housing Authority) 2021.9As with his art and design works, this album is structured with lines, points, and planes. The first part ofthe book depicts the Jing'an Temple, Nanjing Road Commercial Street, and Bund landscape, which connectalong lines to form a classic tourist route; in the second section, 30 historical buildings, designed by L.E.Hudec and scattered all over Shanghai, are presented. At the conclusion of the ICONIC SHANGHAI -When Art Meets Hudec Exhibition, all 30 of the paintings were donated to the Shanghai Urban PlanningExhibition Center; the third part depicts three of the most representative gardens from among the classicalgardens in Shanghai. Zhang's Yu Garden works were selected for the 2013 China Post special postagestamps. Jiading and Nanxiang, two ancient towns, are the destination point for the artist's "nostalgia" andhe poured out his feelings and love in the paintings; the fourth part revisits the grand occasion of the 2010 Shanghai World Expo through the medium of his paintings, and shows the integration of Shanghai urbanculture into the world of the new century. To read about architecture is to read a humanistic story, beginningwith the building exterior and moving inside to the human stories. It is a process of blending vision andemotion. Maybe you can also experience the analysis of the problems of urban development and historicalpreservation. Therefore, Shanghai Sketchbook is not only a pictorial guidebook for tourism and "clockingin", but it also carries the significance of urban culture.

The pen and ink with watercolor medium uses lines as the skeleton and color as the body. It combinesthe characteristics of sketching and watercolor painting. This is the strength of Zhang Anpu's visual art. Heis adept at flexibly using the different properties of hard pen and Chinese brush to describe architecture. Theseemingly random pen lines accurately outline the structural beauty of the buildings, and the perspective isaccurate; the colors, saturated with water, show the "yin-yang" foreground and background of the buildingsin the contrast of cold and warm, leaving a brilliant watercolor wash. In the painting Bund of Shanghai, thepen lines are even hidden to highlight the charm of watercolor painting. In the painting Shanghai DumplingsServed on the Old Street, the color application technique of black-and-white contrast is used to set off thestreet banquet scene of the line drawing. Zhang Anpu basically does not color the characters with an eyetowards ornamentation, but focuses on the vivid posture of the characters. In the use of blank space, he givesfull play to this artistic treatment method of combining virtual reality and "more with less", which is uniquein the field of watercolor painting in Shanghai.

Zhang Anpu used his signature artistic style to record the buildings with different characteristics indifferent times, restoring the charm of Shanghai city of the past century, and letting readers enjoy the beautyof painting and architecture in one volume!

Zhu Guorong (Former Vice Chaiman and Secretary General of Shanghai Artists Assocation)
Zhu Zhirong (Former Director of Xuhui District Housing Authority)
2021.9

目录

壹 · 静安寺 · 南京路 · 外滩 / 007

贰 · 邬达克留给上海的经典老建筑 / 053

叁 · 豫园 · 嘉定 · 南翔 / 115

肆 · 世博园区 / 151

地图上的地点与街道：

- 北京西路
- 南京西路
- 延安中路
- 淮海中路
- 延安西路
- 茂名北路
- 陕西南路
- 西藏中路
- 刘长胜故居
- 宋氏老宅
- 史量才住宅
- 百乐门
- 常德公寓
- 静安寺
- 上海展览中心
- 中福会少年宫
- 蔡元培故居
- 马勒别墅
- 熊佛西楼
- 上海城市规划
- "五卅"纪念碑
- 明天广场
- 上海大剧院
- 中共一大会址

壹

·

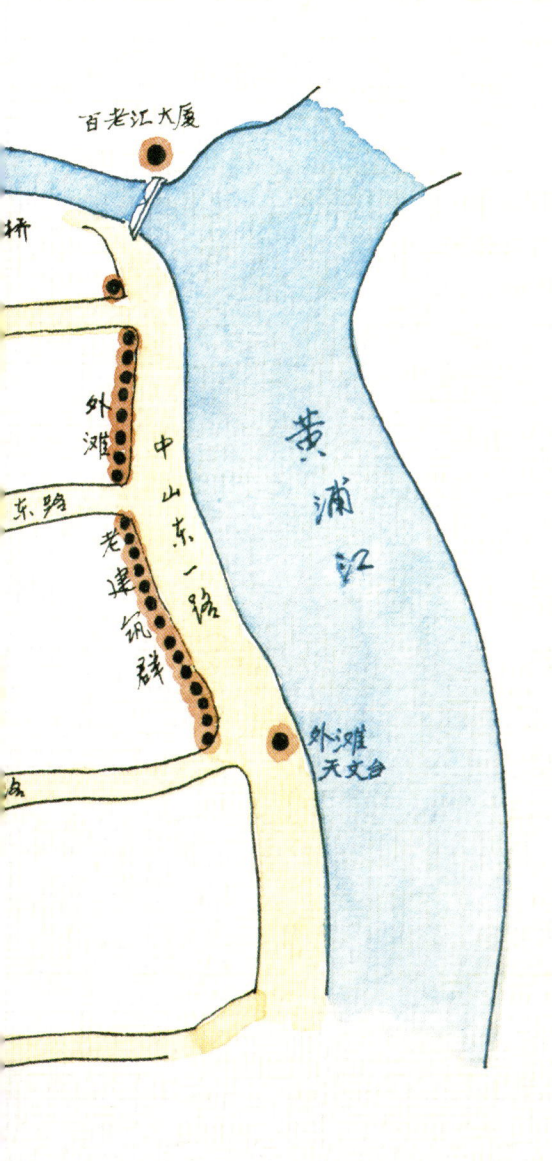

静安寺·南京路·外滩

静安寺

静安寺，这座香火鼎盛的庙宇由名贵楠木建构而成，四周环绕着蜿蜒曲折的亭台楼阁，地处繁华喧闹的都市，周围是林立的现代钢筋水泥高楼大厦，看似突兀，细品之下却又让人感觉到一种闹中取静的和谐。

Jing'an Temple

Jing'an Temple, filled with incense and exuding the aroma of wood, stands amid the hustle and bustle of the downtown. Surrounded by modern skyscrapers of hardened steel and cement, and the twists and turns of pavilions and towers, Jing'an Temple appears seemingly unexpected. But if you savor the atmosphere, then you may just feel a quiet harmony out of the chaos.

静安寺位于静安区南京西路1686号，是上海最古老的佛寺。整个庙宇形成前寺后塔的格局，由大雄宝殿、天王殿、三圣殿三座主要建筑构成，气势宏伟。

张幼朴作于庚子夏日

百乐门

百乐门自开张之日起,便成为上海著名的综合性娱乐场所,号称"东方第一乐府"。在1933年的开张典礼上,时任国民政府上海市市长的吴铁城发表祝词。此后营业期间,张学良、徐志摩等名人经常流连于此,陈香梅与陈纳德的订婚仪式亦在此举行,就连卓别林夫妇访问上海时也曾慕名到访。

The Paramount

Since its opening day, the Paramount has been one of Shanghai's most well-known entertainment venues, and during its heyday was considered the top music and dance hall in the Far East. At the opening ceremony in 1933, Tiecheng Wu, the mayor of Shanghai, personally attended and delivered his blessings. Since then, Xueliang Zhang, Zhimo Xu, and other celebrities have often lingered here. The engagement ceremony of Xiangmei Chen and Nade Chen was held here. Even the Chaplin couple was drawn to the Paramount during their visits to Shanghai.

静安寺百乐门·上了年岁的"老克勒"舞厅

刘长胜故居

愚园路81号是1946年至1949年刘长胜任中共中央上海局副书记时的居住地,也是中共中央上海局的秘密基地之一。该宅是一幢沿街的砖木结构的三层楼房。2004年5月27日,中共上海地下组织斗争史陈列馆暨刘长胜故居正式对社会开放。

Former Residence of Liu Changsheng

No. 81, Yuyuan Road was the residence of Comrade Liu Changsheng when he served as the deputy secretary of the Shanghai Bureau of the CPC Central Committee from 1946 to 1949. It was also one of the secret organs of the Shanghai Bureau of the CPC Central Committee. The house is a three-story building made up of wood and brick along the street. On May 27, 2004, the Exhibition Hall of Shanghai Underground Organization Struggle History of the Communist Party of China & the Former Residence of Liu Changsheng were officially opened to the public.

刘长胜故居位于静安区愚园路81号,这是一幢砖木结构的三层洋房。一九四六年至一九四九年刘长胜同志任中共中央上海局副书记时的居住地,也是中共中央上海局的秘密机关之一。二〇〇〇年五月芒日,这幢楼筑作为中共上海地下组织斗争史陈列馆正式对社会开放。

陆安朴作于庚子夏日

史量才住宅

铜仁路257号，坐落着一幢优雅的花园住宅。该房屋的建筑面积2,494平方米，园地面积2,276平方米，它的主人史量才曾是上海报界巨头、《申报》总经理。史量才住宅建造于1920年左右，从建筑风格看，堪称中西融会的精品，显示了史量才这位报界巨头的独特品位。

The Former Residence of Shi Liangcai

There is an elegant garden residence located at No. 257 Tongren Road that occupies 2,494 square meters of construction area, including 2,276 square meters of building and garden area. Its owner was Shi Liangcai, the Shanghai press tycoon and general manager of Shen Bao. This historic former residence was built around 1920. Its architectural style makes it a work of art that perfectly integrates western and eastern styles.

史量才旧居位于静安区铜仁路257号。建于一九一三年至一九二〇年，为中西结合式的花园住宅。张安朴作于庚子夏月

宋氏老宅

宋氏老宅位于陕西北路369号，建于1908年，是一幢两层半高、四面临空的英国式花园别墅住宅。住宅宽敞明亮，楼前绿树成荫，房屋建筑面积824平方米，园地面积1,218平方米。宋庆龄在这里创办了上海第一个新型的托儿所——中国福利会托儿所。

Song House

Located at No. 369, North Shaanxi Road, Song House was built in 1908 with a two and a half storey. It is a British-style garden villa without walls all sides. The house is spacious and bright, shaded by trees in the front. The building area of the house covers 824 square meters, and the garden area 1,218 square meters. Song Qingling founded the first new type of nursery in Shanghai—the China Welfare Institute Nursery.

宋氏旧居

位于静安区陕西北路三六九号。为村树别墅风格的小洋楼建于一九〇八年，曾是宋庆龄的岳父倪老夫人居其子女的住所，整个庭院风格挺拔，两有气派，回廊佛堂更起、惟静典雅。一九四八年庆龄在此居住，后宋庆龄作为中国福利会北儿物剧会基金会的临时办公地，现列占满宋庆龄基金会办公场所。

砚岛朴作于己亥秋月

熊佛西楼

熊佛西楼位于华山路630号,系由德侨龙特、冯都林等人创立的寓沪德侨社交和娱乐场所,20世纪40年代曾为中央电影公司,1956年起为上海戏剧学院所用,2000年改名为"熊佛西楼"。这里是上海电影人会聚之所,许多电影在此配音、配乐、录歌,一直到1956年上海戏剧学院搬入为止。

Xiong Foxi Building

Xiong Foxi Building, 630 Huashan Road, is a social and entertainment place for Germen living in Shanghai, which was established by German Longte and Feng Dulin. It was the Central Film Company in the 1940s. It has been used by Shanghai Theater Academy since 1956 and renamed as "Xiong Foxi Building" in 2000. This is the gathering place of Shanghai filmmakers. Many films dub, play and record songs here until 1956, Shanghai Theater Academy moved in.

上海戏剧学院熊佛西楼位于静安区华山路630号,是以上海戏剧学院第一任院长熊佛西命名的建筑,它不仅代表了对著名戏剧家熊佛西的怀念,也是上海戏剧学院历史的见证和精神的象征。

张安朴作于庚子秋日

中
国
福
利
会
少
年
宫

中国福利会少年宫成立于1953年6月1日，由中华人民共和国名誉主席宋庆龄亲手创办，是全国第一家少年宫。这里曾经是英籍犹太人嘉道理爵士的住宅。

Children's Palace of the
China Welfare Institute

The Children's Palace of the China Welfare Institute was founded by The Honorary Chairman of The People's Republic of China Soong Ching-ling on June 1st, 1953, as was the first children's palace in China. What is rarely known about the Children's Palace of the China Welfare Institute is that this magnificent building used to be the residence of the British-Jew Sir Ellis Kadoorie.

嘉道理爵士住宅，现为中国福利会少年宫。位于静安区延安西路64号。始建于一九二〇年，因大量使用大理石作建材，被称为"大理石宫"。陆家朴作于庚子夏月

蔡元培故居

这是一幢黄色花园洋房,外形采用较陡的双坡红瓦屋顶,山墙一段露出深色的木构架,墙面镶嵌深灰色卵石,显得亲切而高雅。整幢建筑采用不规则布局,南面有大片草坪,左侧有一间低矮的花房,屋旁两侧植有龙柏、芭蕉、罗汉松、瓜子黄杨和盘槐。整幢房屋占地1,481平方米,建筑面积526平方米。

The Former Residence of Cai Yuanpei

Cai Yuanpei's former residence is a yellow garden villa with a steep double-pitch, gabled, red-tiled roof and dark gray pebble walls. The exposed dark wooden frame stretches on a section of gabled walls, which looks warm and elegant. The building structure has an irregular layout, with a large lawn to its south, a small greenhouse on the left, and cypress, plantains, conifers, evergreens, and Japanese pagoda trees planted on both sides. The whole estate covers an area of 1,481 square meters with a building area of 526 square meters.

蔡元培故居位于静安区华山路303弄16号，是一幢三层英式花园洋房。为蔡元培在上海的最后一处住所，亦是国内保存最完好的一处蔡元培故居。
张安朴作于庚子夏日

上海展览中心

在哈同花园（爱俪园）的废墟上建造的上海展览中心曾经数易其名，从最初的"中苏友好大厦"到20世纪60年代的"上海展览馆"，直至现名。该建筑由苏联建筑艺术家安德列耶夫设计，于1955年3月建成，占地8万平方米，属于俄罗斯风格的建筑。

Shanghai Exhibition Centre

Shanghai Exhibition Centre was built on the site of Silas Hardoon's home, Hardoon Gardens (also called Aili Gardens). Its original name was SinoSoviet Friendship Building, which was changed to Shanghai Exhibition Hallin the 1960s, and to Shanghai Exhibition Centre today. Sino-Soviet Friendship Building was designed by the Soviet architect, Sergey Andreyev. Constructionof the building began in 1954 and was completed in March 1955. The buildingcovers an area of 80,000 square meters and features Russian style.

上海展览中心位于静安区延安中路1000号，建于一九五五年，是上海的代表性建筑之一，也是五十年代上海市建造的首座大型建筑，属于俄罗斯古典主义建筑风格。陆安朴作于庚子夏日

马勒别墅

马勒别墅是一座具有浓郁北欧风情的花园别墅，被誉为上海近代建筑奇迹之一，富有童话世界中的迷幻意境。建成后的别墅的主建筑为三层斯堪的纳维亚式风格建筑，两座主塔高大、挺拔，上开多层小窗，周边建有许多小的尖塔，高低不一的塔尖构成了神秘奇妙的轮廓。整座楼楼面呈现赭红色，造型绮丽；外墙用耐火砖建造，镶嵌彩色瓷砖。一眼望去，别墅宛如童话中色彩斑斓的城堡。

Moller Villa

When arriving at this strong Nordic style garden villa and staying in the charming building which is hailed as one of the modern miracles in Shanghai, you will feel the rich psychedelic mood of Fairy Tales. The villa's main three-floor building is in Scandinavian style, the two main towers are tall and straight, like erect swords, with multi-story small windows. All of the corners of the building are decorated with small steeples and these high and low steeples constitute a mysterious and wonderful outline in beautiful shapes. The entire building is in red ocher, made with firebrick and embedded with colorful ceramic tiles, just like a colorful fairy tale castle at first glance.

马勒别墅位于静安区陕西南路30号。由英籍富商马勒建造，历时九年，二九三六年竣工。主建筑为三层北欧风格建筑，宛如童话世界里的城堡。

张安朴作于庚子夏日

明天广场和上海大剧院

明天广场由世界著名的约翰·波特曼设计师事务所设计，大楼线条硬朗明快，外形十分前卫，像一枚巨型火箭。上海大剧院位于人民广场西部，按照中国古典建筑翘檐式外形设计，屋顶采用两边反翘的白色弧形结构与天空拥抱，如飞翔的白鹤。

Tomorrow Square and Shanghai Grand Theatre

Tomorrow Plaza, designed by the world famous John Portman & Associates architectural design firm, has strong and bright lines creating a very avant-garde appearance that resembles a giant rocket. Located in the west part of People's Square, the Shanghai Grand Theater was designed in the style of Chinese classical architecture. The roof incorporates a white arc structure with both sides tilted up to embrace the sky, like a flying white crane.

明天广场和
上海大剧院
陈安朴作于
壬辰秋日

"五卅"运动纪念碑

位于南京西路、西藏中路交叉口。1925年5月30日，震惊中外的"五卅"运动爆发。它是中国共产党直接领导的以工人阶级为主力军的中国人民反帝爱国运动，标志着大革命高潮的到来。

May 30th Movement Monument

This memorial is located at the intersection of West Nanjing Road and Middle Xizang Road. The May 30th Movement, which sent shock waves through China and foreign countries on May 30, 1925, was an anti-imperialist patriotic movement led by the Communist Party of China and with the Chinese working class as its driving force. The incident marked the arrival of the climax of the great revolution.

位于上海市中心南京西路的大型纪念碑雕塑——五卅运动纪念碑
张罗朴作于壬寅春日

上海城市规划展示馆

上海城市规划展示馆位于人民广场东部，以"城市、人、环境、发展"为主题，展示了上海城市规划建设发展的成就，让人们全面、深入了解上海。

Shanghai Urban Planning Exhibition Center

The Exhibition Center is located in the east side of People's Square. With the theme of "City, People, Environment and Development", it highlights the achievements of Shanghai's urban planning, construction and development initiatives, leaving visitors with a comprehensive and deep understanding of Shanghai.

上海城市规划展示馆位于人民大道100号，要了解上海从这里开始。张安朴作于丙申秋日

南京东路是上海著名的商业步行街，东起外滩，西至西藏中路，全长1,599米。图片（右）为南京东路的和平饭店。

南京东路

East Nanjing Road

Shanghai's famous commercial pedestrian street runs from the Bund on the east to Middle Xizhang Road on the west, for a total length of 1,599 meters. This painting shows the Peace Hotel (on the right) situated at the entrance to East Nanjing Road.

上海南京东路外滩

上海南京东路是全国著名的商业步行街，东起外滩，西至西藏中路，全长一千五百九十九米。位于这里的建筑是和平饭店北楼与和平饭店南楼，均是全国重点文物保护建筑。

张安朴作于壬寅春日

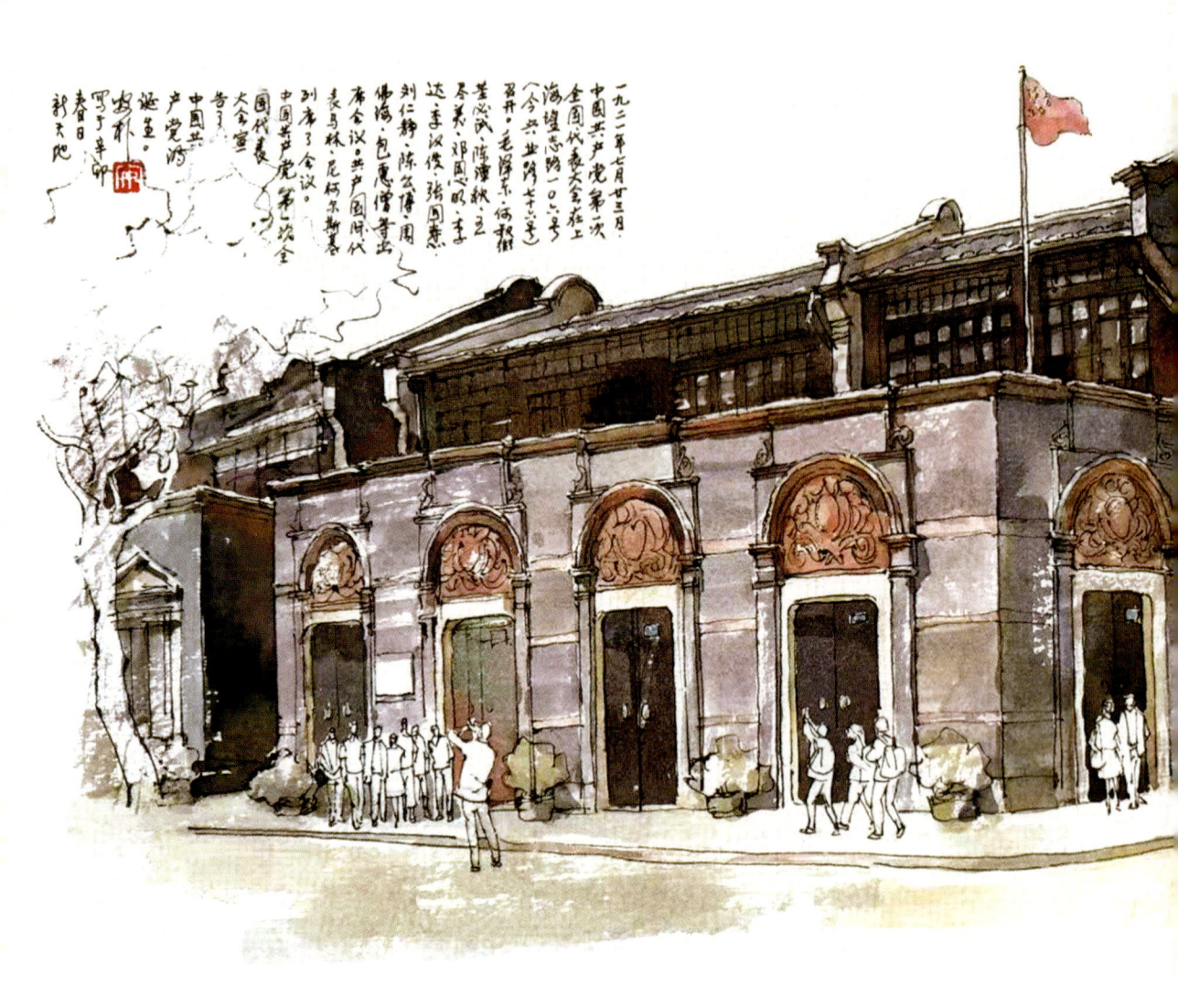

一九二一年七月廿三日，中国共产党第一次全国代表大会在上海望志路一〇六号（今兴业路七十六号）召开。毛泽东、何叔衡、董必武、陈潭秋、王尽美、邓恩铭、李达、李汉俊、张国焘、刘仁静、陈公博、周佛海、包惠僧等出席会议。共产国际代表马林、尼柯尔斯基列席了会议。中国共产党第一次全国代表大会宣告，中国共产党诞生。

牧朴写于辛卯春日 郭天地

中国共产党第一次全国代表大会会址纪念馆

纪念馆位于上海新天地，是一幢沿街的旧式石库门住宅建筑。中国共产党第一次全国代表大会于1921年7月23日至7月30日在该楼一楼客厅举行。此建筑在1952年后成为纪念馆。

Memorial Museum at the Site of the First National Congress of the Communist Party of China

This memorial museum is located in Xintiandi, Shanghai, in a historical Shikumen residence building just beside the street. The First National Congress of the Communist Party of China was held in the first floor living room from July 23 to July 30, 1921. The building has been established as a memorial site since 1952.

浙江路桥

浙江路桥，作为苏州河上形制最奇特的一座桥，是世界上仅存的几座鱼腹式钢桁架结构桥梁之一，已被列为上海市近代优秀建筑。

Zhejiang Road Bridge

Zhejiang Road Bridge, as a peculiarly-shaped bridge on Suzhou Creek, is one of the few remaining fish-bellied steel truss bridges in the world, listed as an outstanding morden building in Shanghai.

浙江路桥横跨苏州河，坐落于浙江中路与浙江北路连接处，是罕见的鱼腹式简支钢桁架桥。建于一九〇六年，为上海市文物保护单位。

陆发朴作于庚子夏日

原百老汇大厦原为供来华外国人使用的旅馆兼公寓，由公和洋行设计，新仁记营造厂施工。这一座铝钢框架结构的二十二层大楼，于1934年建成，糅合了装饰艺术派与美国现代高层建筑风格。1951年该建筑更名为上海大厦。

原百老汇大厦

Formerly Broadway Mansions

Previously used as a hotel and an apartment house for foreigners in China, the house, whose construction was designed by Palmer & Turner, and built by the contractor XinRenJi. Completed in 1934, the twenty-two-storey building with an aluminum-steel structure is an example of decorative art and modern American skyscraper style. In 1951, the building was renamed "Shanghai Mansions."

041

原沙逊大厦、华懋饭店

今和平饭店，位于外滩20号。1929年由沙逊家族投资，新沙逊洋行第三任大班英籍犹太人爱利斯·维多克·沙逊主持建造，英商公和洋行设计，华商新仁记营造厂承建。建筑整体属装饰艺术派风格，采用钢框架结构。

Formerly Sassoon House or Cathay Hotel

It is Peace Hotel nowadays, which located at No. 20 on the Bund. This building, whose construction was funded by the Sassoons in 1929 and supervised by British Jew, Sir Alice Victor Sassoon, the third president of E. D. Sassoon & Co., was designed by Palmer & Turner of the UK and built by the contractor XinRenJi. On the whole, the building, with a steel structure, is an example of decorative art.

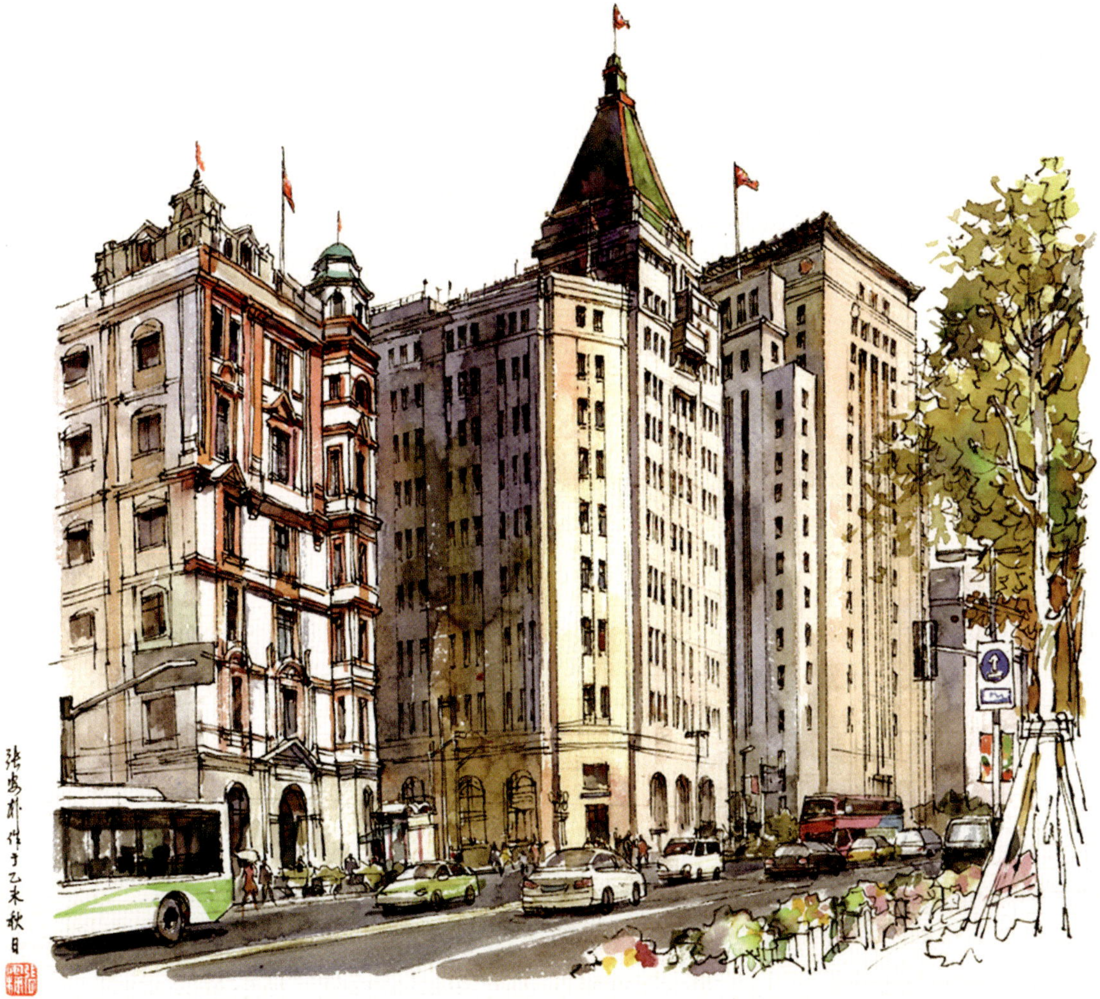

原汇丰银行大楼

今上海浦东发展银行大楼，1923年竣工，属于新古典主义风格的建筑。采用钢框架结构，以砖墙填充，外墙以花岗石贴面。占地面积5,500平方米，是外滩占地最大、门面最宽的建筑。

Formerly Hong Kong & Shanghai Banking Corporation Building

It is Shanghai Pudong Development Bank Building nowadays. The building, completed in 1923, is an example of Neo-Renaissance architecture. Its steel structure is propped up with brick walls. The external walls are veneered with granite. The building occupies the largest area (i.e., 5,500 square meters) with the most spacious facade among all the buildings on the Bund.

上海外滩建筑群是上海百年历史沧桑的见证。刚健雄伟、雍容、华贵的气势，是新老上海的象征之一。
陈安朴作于癸卯春日

原亚细亚大楼

今上海久事国际艺术中心，1915年竣工，初名"麦边大楼"，1917年改名为"亚细亚大楼"。建筑整体属新古典主义风格，为钢筋混凝土框架结构，高八层。

Formerly Asia Building

It is Shanghai Jiushi International Art Center nowadays. This house, completed in 1915, was initially named "Mcbain Building." In 1917, the Asiatic Petroleum Company of the UK acquired it with a new name, "Asia Building." The eight-storey building with a reinforced concrete frame is an example of Neo-Renaissance architecture.

047

外滩信号台

外滩信号台位于外滩、延安东路路口，始建于1907年，是上海最早的报时、报天气信息机构。

Bund Signal Station

Founded in 1907, the Signal Station is located at the intersection of the Bund and East Yan'an Road and is the earliest time and weather information facility in Shanghai.

外滩信号台塔称外滩天文台,位于中山东二路一号,一九零七年建立圆柱形气象信号台,总高50米,一九九三年被碧悠大厦边移动22米,立于现址。张安朴作于丁酉春日

外滩全景

外滩北起外白渡桥，南抵金陵东路，全长约1.5公里。临江而立、巍峨参差的外滩万国建筑博览群，是上海百年历史沧桑的见证。

The Bund

The 1.5-km-long Bund begins at the Waibaidu Bridge (formerly Garden Bridge) in the north and ends at Jinling Road East in the south. Buildings of various architectural styles stand on the western bank of the Huangpu River, as witnesses of the hundred-year vicissitudes of Shanghai.

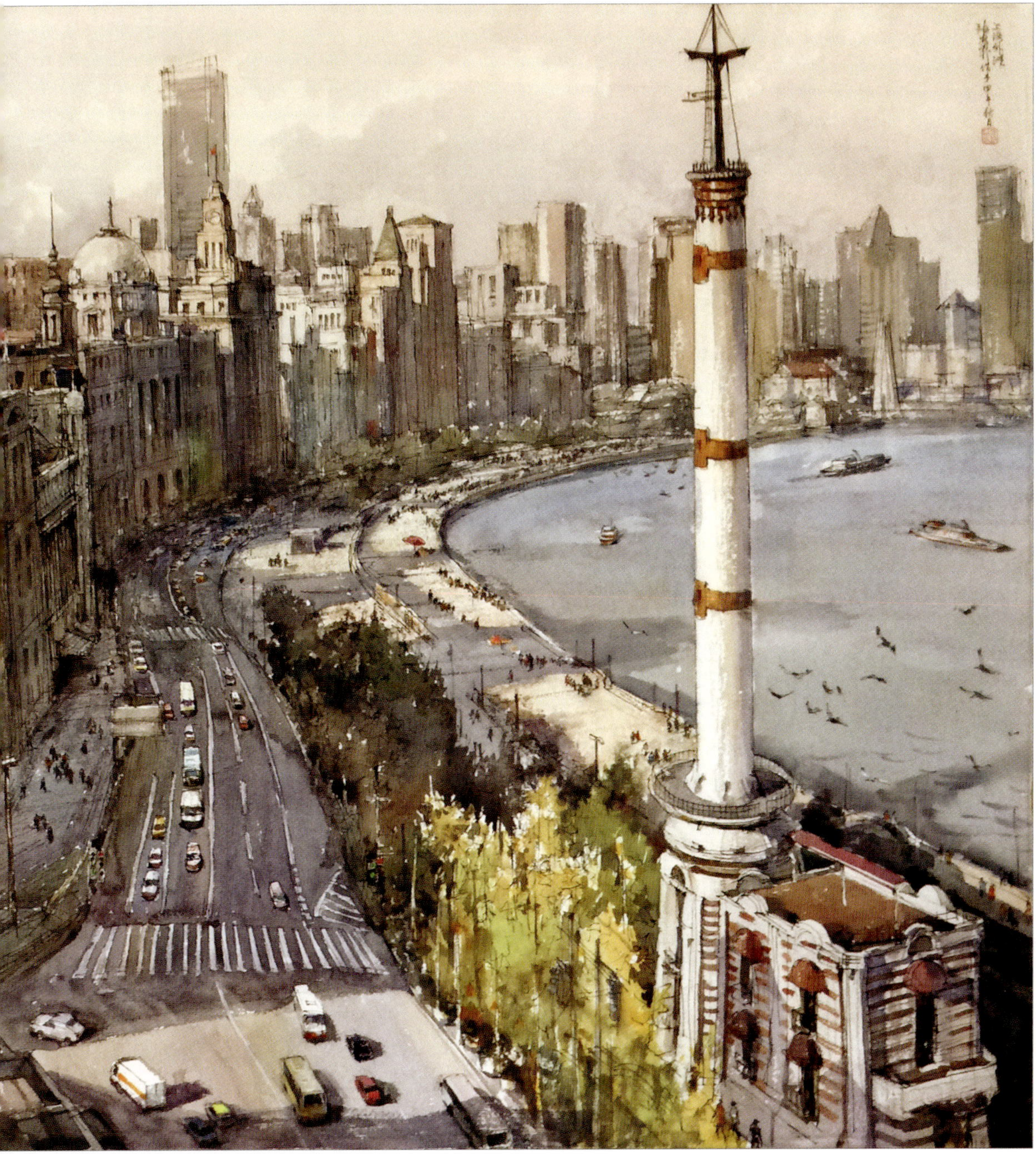

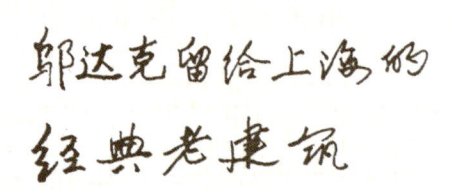

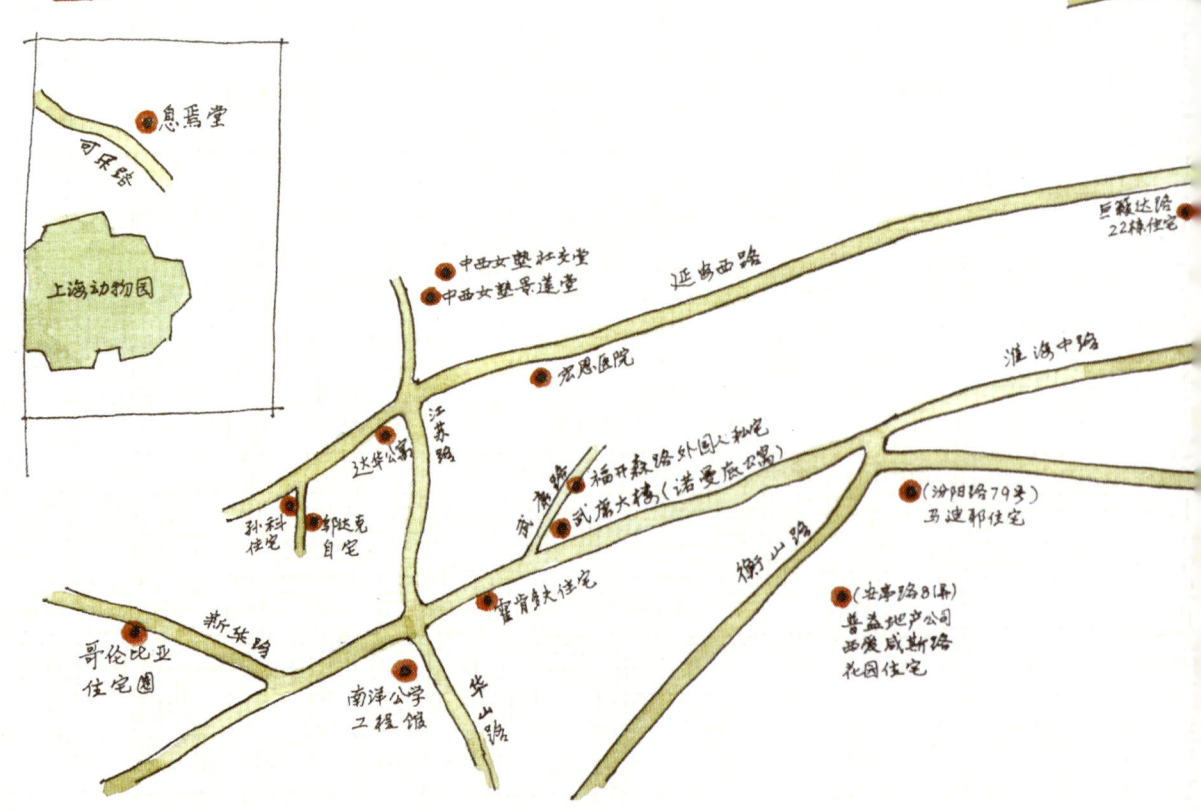

邬达克留给上海的经典老建筑

贰

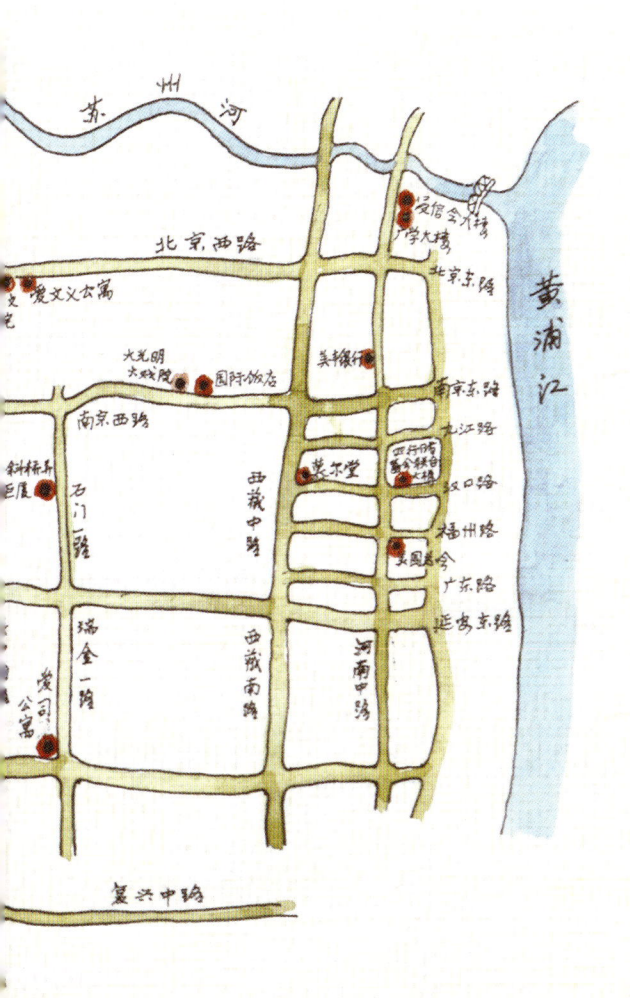

·

邬达克留给上海的

经典老建筑

吴同文住宅

吴同文住宅（今上海市城市规划设计研究院，铜仁路333号），1935年9月设计，1938年7月竣工。占地面积22,200平方米，建筑面积2,000余平方米，四层，钢筋混凝土框架结构。因整幢楼建筑外墙贴绿色釉面砖，老上海人习惯称其为"绿房子"。

该建筑是邬达克现代风格的重要代表作，也是上海近代花园洋房中的经典。建筑总体布局紧凑合理，与基地完美结合。该建筑曾被当时的《中国月刊》誉为"整个远东地区最大最豪华的住宅之一"。

D.V.Woo's Residence

D.V.Woo's Residence, today as The Shanghai Urban Planning Design Institute, is located at No. 333 Tongren Road. Designed in September 1935 and completed in July 1938, the four-storey reinforced concrete structure covers 22,200 square meters of site area and over 2,000 square meters of building area. Because of the surface covered with green glazed tiles, it is commonly known as "the Green House" by people of old Shanghai.

This residence is one of Hudec's masterpieces which integrated a modern architecture with Art Deco style. The layout of the house was compact and China Journal valued it as "one of the most spacious and luxurious residences in the Far East".

绿房子位于铜仁路333号的吴同文住宅，是邬达克设计的现代新风格建筑，于一九三八年竣工。因整幢建筑外墙贴绿色釉面砖，被称为"绿房子"。

张安朴作于丙申夏日

国际饭店

国际饭店（南京西路170号）建筑外形仿美国早期摩天楼形式，立面强调垂直线条，层层收进直达顶端，高耸且稳定的外部轮廓，尤其是十五层以上呈阶梯状的塔楼，表现出美国装饰艺术派的典型特征。这座从落成至1983年称雄半个世纪"上海之巅"的建筑，是邬达克现代派思想和装饰艺术风格的代表作，是继大光明影院后更为大胆的探索和创造。借助当时的进步科技、战后涌向房地产业的庞大的国际资金以及业主经济实力的共生效应，国际饭店达到当时远东高层建筑设计和施工的最高水平。

Park Hotel

Park Hotel is located at No. 170 West Nanjing Road. Taking early American skyscrapers in New York and Chicago as a source of inspiration, J.S.S Building was designed in a style of modern Art Deco with long vertical stripes shrinking layer upon layer on the elevations and creating a setback skyline at the top of the tower above the 15th floor. J.S.S Building has been reputed as the tallest landmark in Shanghai for nearly half a century. It is Hudec's masterpiece of modernism and Art Deco by virtue of rapid development in both financial investment in real estate and building technology in Shanghai in 1930s, the design and the construction of the Park Hotel reached the highest level of skyscrapers in the Far East in 1930s, and keep the record for more than 50 years.

国际饭店,位于上海南京西路一百七十号,是邬达克设计的代表作品。张安朴作于丙申夏日

国际饭店高83米,是称雄半个世纪的"上海之巅"。安朴题记

刘吉生住宅

刘吉生住宅（今上海市作家协会，巨鹿路675号）是仿欧洲古典希腊式建筑，造型宏伟庄重，气派非凡。该建筑因按希腊神话中爱神丘比特和普绪赫的故事而设计，又被称作"爱神花园"。住宅的许多细部特征，在英国著名学院派画家莱顿的名画《普绪赫洗浴》中都能找到原型。

Liu Jisheng's Residence

Liu Jisheng's Residence, today as Shanghai Writers' Association, is located at No. 675 Julu Road. Imitating the Greek architectures in Europe, the garden house is a magnificent project. Inspired by the stories of the Cupid, the God of Love, and his wife Princess Psyche, the garden was called later by people as the "Cupid's Garden". The archetype of many detailed characteristics of the house can be found in the well-known drawing "the Bath of Psyche" by the famous English academism artist, Frederic Leighton.

刘吉生住宅位于巨鹿路675号，邬达克于一九二七年设计，住宅为仿欧洲古典希腊建筑，极具浪漫气息，被称为"爱神花园"。现为上海市作家协会办公楼。张安朴作于丙申夏日

中西女塾景莲堂

中西女塾景莲堂（今市三女中五四大楼，江苏路155号）是一幢美国学院派哥特式样的教学楼，立面简洁朴素，外墙面用水泥砂浆仿石材装饰，屋顶用的是双坡硬山红瓦。底层开长方形窗，二至三层则为方形窗，加之屋顶上三组整齐排列的老虎窗，整个外形庄重大方又不失活泼生动。

McGregor Hall, McTyeire School for Girls

McGregor Hall (McTyeire School for Girls) today as Shanghai No. 3 Girl's Middle School, is located at No. 155 Jiangsu Road. As a Christian girls' school, the McTyeire School was founded in 1890. As a teaching building of American gothic style, the McGregor Hall has a simple facade, ashlar brick decorating external walls by cement mortar, and a flush gable roof consisting of red tiles. The shapes of windows vary from rectangular on the ground floor, to square on the second and the third floor to the dormers, regularly arranged on the roof, making the whole structure, substantial and appealing.

中西女塾景莲堂,位于江苏路155号,今为市三女中教学楼。五大楼由邬达克设计,一九三五年建造。景莲堂是一幢美国学院派哥特式样的教学楼,一九九四年被列为上海市优秀历史建筑。

张安朴作于丙申夏日

慕尔堂

慕尔堂（今沐恩堂，西藏中路316号）的建成，表明邬达克对设计手法及古典建筑语汇的把握已日趋成熟和自如，可以灵活处理复杂的功能要求、基地条件和道路情况，也开始对营造不同的空间氛围和特色做更多的理性思考。

Moore Memorial Church

Moore Memorial Church is located at No. 316 Middle Xizang Road. The completion of Moore Memorial Church witnessed Hudec's maturity in the free application of classical architectural styles. He was able to deal with complicated functional requirements, foundations and the surrounding street-scape, and to cast rational thought toward the creation of an environment of a special character.

上海的邬达克建筑—沐恩堂 张安朴作于丙申夏日

沐恩堂，位于西藏中路三百十六号，建于一九三一年，为上海市优秀历史建筑。安朴题记

广学大楼

广学大楼（虎丘路128号），原为广学会（1887年英国长老会传教士创建同文书会，1892年改名广学会，是当时中国最大的出版机构）所有。它与浸信会大楼在平面上贯通，是两幢连体姐妹楼。广学大楼立面垂直高耸，气势宏伟，行人仰止。

Christian Literature Society Building

Building is located at No. 128 Huqiu Road. It was originally owned by the Christian Literature Society, a Society for the Diffusion of Christian and General Knowledge among the Chinese, which had been established by the Presbyterian missionaries from England in 1887 and changed its name to the Christian Literature Society in 1892, becoming the largest publishing organization in China at that time. The building is linked with the Baptist Publication Society Building in the layout. The two are twin buildings.

广学会大楼,位于虎丘路128号,邬达克设计,一九三二年竣工。大楼整体为现代派装饰艺术风格,立面垂直高耸,气势宏伟。一九九〇年被列为上海市优秀历史建筑。张安朴作于丙申夏日

南洋公学工程馆

南洋公学创办于1898年，1921年改名为交通大学。该工程馆（今上海交通大学工程馆，华山路1954号）为现代装饰艺术风格，外墙镶深褐色面砖，石材壁柱凸出墙体呈锯齿状，强调竖线条。朝向内院的北立面入口是三个尖券门，透出邬达克对哥特式风格的喜爱与垂青。

Engineering Building in Nan Yang Public School

Today as Engineering Building in Shanghai Jiao Tong University, is located at No. 1954 Huashan Road. Normal College of Nan Yang Public School is established in 1986 and changed its name into Shanghai Jiaotong University in 1921. Designed in Modern Art Deco style, the facade is covered with dark brown facing tiles with stone pilasters projecting beyond the walls to emphasize the vertical lines. Three pointed-arches on the north entrance facing the garden show Hudec's preference for Gothic style.

南洋公学工程馆（今交通大学工程馆），位于华山路一九五〇号，邬达克设计，一九三二年竣工。在这一回字形教学楼里培养出许多杰出人才。张安朴作于丙申夏日

普益地产公司西爱咸斯路花园住宅

住宅区分别位于永嘉路563、615、623号，安亭路41弄16、18号，81弄2、4号，建于1938年，是西班牙风格的联立式花园住宅。走近这幢联立式住宅可以发现，从住宅的布局到各个细部，无不体现它是一幢西班牙风格非常显著而统一的近代建筑。住宅左右两个单元完全对称，围成半私密三合院空间。住宅在其他细节上西班牙风格的建筑特征也相当显著，如螺旋柱、无线脚的门窗券、红色筒瓦的缓坡顶、淡黄色粉墙等等。两个单元的主入口门框装饰相当惹目，门头曲线强劲流畅，与三合院端部的山墙曲线相呼应，透着西班牙式的巴洛克神韵。

Garden Villas on Route Sieyes for Asia Realty Co.

This Spanish style garden row house was built in 1938, is located at No. 563&615&623 Yongjia Road, No. 16&18 (Lane 41) and No. 2&4 (Lane 81) Anting Road. Coming near the residence, you'll find it's a modern building with remarkable and uniform Spanish styles from its layout and every detail. The two units on the right and left sides of this residence are completely symmetric, forming a semi-private space of a courtyard enclosed in three directions. The Spanish styles are remarkably reflected in other details of this residence, such as the wreathed columns, doors and windows without architraves, a gentle slope roof with red semicircle-shaped tiles and light yellow walls. The doorframe decorations in the main entrances of the two units are pretty striking. The curves on door heads are strong and fluent, corresponding with the curves of the gable walls on the top of the courtyard and revealing a Spanish Baroque style.

普益地产西爱成斯路路花园住宅位于今天的安亭路81弄2号，是邬达克建筑作品。于一九三五年至一九三六年建造。西班牙式，平面为三边围合式。红色筒瓦坡屋面，黄色水泥墙面。

张安朴 作于丙申秋日

马迪耶住宅

马迪耶住宅（今上海工艺美术博物馆，汾阳路79号）是一幢法国文艺复兴式建筑，建筑面积1,496平方米。外墙为白色，故称"海上小白宫"。

Madier House

The villa today as Shanghai Museum of Arts and Crafts, is located at No. 79 Fenyang Road. It is a French Renaissance building of brick and concrete structure that covers a building area of 1,496 m^2. With white external walls, it's called "Little White House of Shanghai."

汾阳路79号，是一幢古典主义的白色府邸，由邬达克设计，被列为上海优秀近代建筑。现为上海工艺美术博物馆。

张安朴作于丙申中秋日

何东住宅

何东住宅（今上海辞书出版社旧址，陕西北路457号）是邬达克初到上海就职于克利洋行时的作品。精美的古典主义表现手法、中西结合的室内外布局，已展现出他对流行的把握和对中西文化理解的不凡能力。

Ho Tung's Residence

The residence today as Shanghai Lexicographical Publishing House, is located at No. 457 North Shaanxi Road. It was one of Hudec's early works when he was employed in R. A. Curry's firm. The indoor and outdoor layout that combined the Chinese and Western styles are evidence of Hudec's grasp of a popular trend, his careful consideration of the owner and a superior understanding of Chinese and Western cultures.

何东住宅位于陕西北路457号，今为上海辞书出版社。邬达克早期的设计作品，一九二六年竣工。二层钢筋混凝土结构，带补古典主义风格。二○○五年被列为上海优秀历史建筑。张安朴作于丙申秋月斜桥弄巨厦（今上海公惠医院，石门一路315弄6号）是一座规模宏大的花园洋房，外形以西班牙风格为主调，兼有其他风格的建筑元素。人们从设计中可以看到邬达克对不同阶段、不同地域建筑风格的包容和认同。

P.C. Woo's Residence

This façade today as Shanghai Gonghui Hospital, is located at No. 6 Lane 315 Shimen Road No. 1. It was composed of typical Spanish architectural elements, such as Spanish roof tiles, open loggia, spiral columns and cast iron railings.

斜桥弄巨厦（今为公惠医院），位于石门一路313弄6号，邬达克于一九三一年设计，次年春竣工。这是一座规模宏大的花园洋房，外形以西班牙式为主而建筑风格。二〇一五年被列为上海优秀历史建筑。
张安朴作于丙申秋日

哥伦比亚住宅圈

邬达克在哥伦比亚住宅圈（今私人住宅区，新华路119、155、185、211、236、248、276、294、329号）设计了多幢花园洋房，式样有英国、美国、荷兰、意大利、西班牙等多种风格。这些风格迥异、美轮美奂的住宅，被称作"哥伦比亚住宅圈内的精华建筑群体"。

Columbia Circle

This area today as private residential, is located at No. 119&155&185&211&236&248&276&294&329 Xinhua Road. Hudec designed several garden residences here in English American, Dutch, Italian, Spanish and many other styles. They were termed the "Essential Architecture in the Columbia Circle".

哥伦比亚住宅圈，包括新华路119弄、155弄、185弄、211弄和329弄，时称"外国弄堂"。邬达克于一九六年设计，一九三〇年建造。此图为329弄17号，为西班牙式建筑，上海棉纺业巨子薛福生曾在此居住。二〇〇五年被列为上海优秀历史建筑。

张安朴写于丙申秋月

孙科住宅

孙科住宅（延安西路1262号）于1929年设计，1931年建造。建筑原是邬达克为自己设计的住宅。因建造慕尔堂项目遇到麻烦，得到孙中山之子孙科相助，邬达克因为感恩低价转让给了孙科，自己则在马路对面另建新居（即今邬达克旧居）。该住宅以西班牙和意大利文艺复兴式建筑风格为主，兼融其他多种风格元素。

Sun Ke's Residence

Sun Ke's Residence is located at No. 1262 West Yan'an Road. It was designed in 1929 and constructed in 1931. The dwelling eclectically combines various architectural elements into an overall Spanish and Italian Renaissance style.

孙科住宅位于番禺路60号，一九三一年建造。一九二九年设计，原为邬达克建造自用，后转让给孙科。住宅糅渡了和意大利文艺复兴式建筑风格为主。一九八一年被列为上海市优秀历史建筑。张安朴作于丙申夏末。

美丰银行大楼

美丰银行大楼（今强农大楼，河南中路521—529号）于1924年建成完工，为邬达克早期作品。这栋楼为现代建筑风格，顶层为白色水泥墙面，一、二、三层为耐火砖，墙间有水泥线脚，女儿墙有线脚装饰，转角处理成弧形，一层层高较高，黑色钢窗，窗扇有井字装饰，内部小天井极有特点，空间宜人。大楼基础为条形砖，承重砖墙为黏土砖，混凝土楼板，石棉瓦坡屋面，白铁屋脊。外墙采用清水墙饰面，楼内为混凝土楼梯、铁栏杆、木扶手。

American Oriental Banking Corporation

The building of American Oriental Banking Corporation is an early work of Laszlo Hudec completed in 1924, today as Qiangnong Building, located at No. 521—529 Middle Henan Road. It is of modern architectural style. The top storey presents the white cemented walls. The first, second and third storeys are built with firebricks with cement architraves between walls and on parapets, and the corners are made arc-shaped. The first storey is of rather large height with black steel windows whose sashes are decorated by grids. The small courtyard inside is very distinctive for its pleasant space. With lath bricks as its foundation, this building features clay-brick load-bearing walls, concrete floors, asbestos-tile slope roof and galvanized iron ridges. The dry wall facing is adopted for the outer walls. Stairs inside are all concrete stairs with wooden handrails and iron bars.

美丰银行大楼位于河南中路521—529号和宁波路180号，是邬达克早期设计的作品，建于一九二〇年。二〇〇五年被列为上海优秀历史建筑。张安朴作于丙申夏日

邬达克自宅

邬达克自宅（今邬达克纪念馆，番禺路129号）为英国乡村风格，陡峭的石板瓦双坡屋顶占立面高度近一半，两端是高耸的砖砌烟囱，南立面设计对称的山墙造型。底层采用红色清水砖墙，二层以上则是白墙，深色木结构露明，门窗套为粗粝石质。底层为哥特式三连列窗和圆拱形大门，二层是折线型凸窗。

Hudec's Residence

The house today as L.E.Hudec Memorial Hall, is located at No. 129 Fanyu Road. It was in English country style, and the steep sloping roof, which was covered with slate tiles, was close to half of the elevation height. Two towering brick chimneys stood at both ends of the house, with symmetrical pediments in the South facade. Red brick walls were used for the ground floor, and a white wall for the second floor with exposed dark wood frames. Three sets of Gothic windows and a round arch door were on the ground floor.

邬达克自宅位于番禺路129号，建于一九三〇年。这座自宅是典型的英国乡村风格，邬达克在此居住了六年。该建筑二〇〇五年被列为上海优秀历史建筑。

张安朴作于丙申秋月

息焉堂（可乐路17号）仿拜占庭建筑样式，强调拱、券、穹隆等形式特征。它是上海罕见的拜占庭风格的教堂，颇具东欧风情，反映出邬达克对不同风格建筑的准确把握，体现了他的个人设计偏好及地域情结。

Sieh Yih Chapel

The church is located at No. 17 Kele Road. It imitates the structure of Byzantine architecture, emphasizing basic characteristics such as vaults, arches and domes. Radically different from Hudec's more traditional structures in Shanghai, this Byzantine church shows Hudec's accurate grasp of different architectural styles. And we can take a glimpse of his individual preference and regional complex of design.

息焉堂是邬达克设计的天主教堂，位于可乐路一号，始建于一九二五年。息焉堂是上海罕见的拜占庭风格教堂，颇具东欧风情。一九九四年被列为上海市优秀历史建筑。

张鸣朴作于丙申夏末

巨籁达路22栋住宅

巨籁达路22栋住宅（今为私人住宅，巨鹿路852弄1—8号、10号、868—892号）是英商亚细亚火油公司投资建造的独立式花园住宅，共22栋，沿街整齐排列，皆坐北朝南，砖木结构假三层。房屋南面有大花园，南立面中部呈半圆形突出，上设三连式拱券窗。

22 Residences on Route Ratard

These are detached-houses today as private residential, is located at No. 1—8&10&868—892 Lane 852 Julu Road. It was invested and built by Asiatic Petroleum Company from Britain, consisting of 22 buildings lined up neatly along the street facing south with three created layers of masonry-timber structure. A big garden lies in the south, and the middle of southern wall sticks out in the shape of a half circle with three arched windows.

巨籁达路22栋住宅是邬达克早期设计的作品，位于今天的巨鹿路852弄及常熟路附近的花园住宅。住宅群整体构成统一而丰富的外观，每栋别墅前还有大片草坪。一九九九年这群别墅被分别列入上海市优秀历史建筑保护名录。

丙申秋日张安朴作于静安

霍肯多夫住宅

霍肯多夫住宅（淮海中路1893号）于2015年被列为上海市优秀历史建筑。

Huckendoff Residence

This Residence was listed into Shanghai Outstanding Historical Buildings in 2015. It is located at No. 1893 Middle Huaihai Road.

霍肯多夫住宅为邬达克设计的作品，建造于上世纪廿年代初。位于今天的淮海中路1893号。二〇一五年被列为上海优秀历史建筑。张安朴作于丙申中秋日

福开森路外国人私宅

福开森路外国人私宅（武康路129号）由邬达克于1929年设计，砖木结构，西班牙风格，建筑平面近似梯形。2015年被列为上海市优秀历史建筑。

D. Tirinnanzi Residence on Route Ferguson

This building is located at No. 129 Wukang Road. It was designed by Hudec in 1929, brick and wood structure in Spanish style, with a floor plan resembling a trapezoid. It was listed into Shanghai Outstanding Historical Buildings in 2015.

武康路129号，昔日称之"福开森路外国人私宅"，为邬达克设计的西班牙式住宅，二〇一五年被列为上海优秀历史建筑。

丙申秋日张安朴作于

美国花旗总会（今上海高级人民法院，福州路209号）是邬达克在克利洋行时期最有影响的代表作。他偏好用深色面砖、白色大理石等装饰材料，这影响到他日后设计的多个作品，也开启了20世纪30年代上海建筑中广泛使用棕色耐火砖作外墙装饰的先声。

美国花旗总会

The American Club

The building could be considered as an influential building, in particular, for its use of dark brown refractory bricks as facing tiles on the external walls, which became popular in Shanghai in the 1930s. Today as Shanghai High People's Court, is located at No. 209 Fuzhou Road.

美国花旗总会美国花旗总会，位于福州路209号，是邬达克于上世纪二十年代的设计作品，采用深色面砖和白色大理石材料，是邬达克建筑的特色之一。
张安朴作于丙申夏日

浸信会大楼

浸信会大楼（今真光大楼，圆明园路209号）的设计，反映出邬达克对哥特样式的喜爱。建筑注重整体形象的几何装饰性，以及用面砖饰墙面的个人风格的形成，标志着他正逐步转变为一位追求建筑新风格的先锋建筑师。

China Baptist Publication Society Building

The design of this building reveals the formation of Hudec's individual architectural style, with a preference for Gothic Revival architectural features, particularly notable in the rich brickwork facade. It also witnesses his growth step by step as an eclectic architect able to work in a variety of architectural styles. It is located at No. 209 Yuanmingyuan Road.

真光大楼位于圆明园路209号。邬达克于一九三〇年设计，两年后竣工。大楼坐西朝东，主面为锐角状坚线条装饰，褐色面砖，体现出邬达克建筑的特征，张扬朴作于丙申夏日

中西女塾社交堂

中西女塾社交堂（今市三女中五一大楼，江苏路155号）呈马蹄形，南北对称。底楼正中是小礼堂，当时被称为"海涵堂"。礼堂左右两边是走廊，分别通向宿舍区的楼梯。十根白色的水泥柱子牢牢地支撑着走廊柔和优美的弧线托起的尖顶。四楼的22个老虎天窗也是尖顶红瓦。

Social Hall, McTyeire School for Girls

The southern and northern winds of the building form a symmetric U-shape. There is a small auditorium in the middle of the ground floor called "Haihan Hall" in the past. On the right and left sides of the hall are corridors separately leading to stairs of the dormitory area. Ten white cement pillars firmly support the roof of the corridors. The 22 dormer windows on the fourth floor also have pinnacles and red tiles. Today as a building in Shanghai No. 3 Girls' Middle School, is located at No. 155 Jiangsu Road.

中西女塾社交堂，位于江苏路155号，今为市三女中五大楼。邬达克于一九二二年设计，一九三二年建成。如今与三世中的五一大楼和二四大楼一起被列为上海市优秀历史建筑。
张泓林作于丁酉春日

爱文义公寓

爱文义公寓（今联合公寓，北京西路1341—1383号、铜仁路304—330号、南阳路208—228号）包括两幢四层公寓楼、19个车库及一个锅炉房，沿街还有13个两层的中式店铺，属于20世纪30年代上海公寓建筑"周边式"类型。公寓为现代派风格并带有装饰艺术风格的痕迹，外立面简洁明快，用浅红色面砖饰面，横向以水刷石做水平线条。楼梯间则做竖向构图，建筑沿原道路形状在西侧形成一流畅的弧面。

Avenue Apartments

The residential complex was a typical "surrounding" apartment structure in Shanghai in the 1930s, including two four-floor apartments with 13 two-storey Chinese shops along the street to the north of the site, 19 garages and one boiler room. Being of the modernist style with traces of Art Deco, the simple facade of the apartment was coated with facing tiles in light red, and decorated by the horizontal lines of granite plaster. The vertical stair well was in the west side. Today as United Apartments, is located at No. 1341—1383 West Beijing Road, No. 304—330 Tongren Road, No. 208—228 Nanyang Road.

爱文义公寓即抑今的联华公寓，位于北京西路与铜仁路的丁字路口两侧。邬达克于一九三一年设计次年竣工建筑外观为现代风格，立面简洁，为红色清水砖墙，形成流畅的飘西〇二九九年被列为上海市优秀历史建筑。
张安朴作于丁酉夫相月

宏恩医院

宏恩医院（今华东医院1号楼，延安西路211号）为邬达克独立开业初期所设计，延续了他在克利洋行时期的古典风格，立面为意大利文艺复兴风格。宏恩医院的设计体现出邬达克严谨、扎实的古典主义功底，以及处理功能复杂的公共建筑的特殊能力。

Country Hospital

The building was designed by Hudec in the initial stage of his own firm, which maintained the neoclassical style he used in R. A. Curry. The design of Country Hospital represents Hudec's precise and profound skills in designing classical buildings and his special ability to deal with sophisticated public buildings. Today as No. 1 building in Huadong Hospital, is located at No. 211 West Yan'an Road.

宏恩医院由邬达克设计，为抗今今的华东医院的老楼，位于延安西路221号。一九二二年建造。宏恩医院的设计体现出邬达克严谨扎实的古典主义功底和处理功能复杂的公共建筑的特殊能力。一九九一年被列为上海市优秀历史建筑。张安朴作 丁酉春日

爱司公寓

爱司公寓（今瑞金公寓，淮海中路758号）采用法国文艺复兴风格，立面有明显的横三段划分，裙部饰深褐色，稳重端庄，上部腰线和檐口线则又十分轻巧，双重檐口间饰半圆券的三连窗与水平线条相连。立面间隔布置有强调竖向构图的凸形窗，凸显公寓的高大。凸形窗设计竹节状细圆柱，窗肚墙作框，既显华丽又可使住户视线开阔。

Estrella Apartments

Estrella Apartments today as Ruijin Apartments, is located at No. 758 Middle Huaihai Road. In this period, most of Hudec's works were designed in the French Renaissance beaux-arts manner. The facade of the apartment had distinct horizontal triadic divisions and the skirt was decorated in dark brown. Meanwhile, the upper waist and the cornice line were light. The triple casement window ornamented with semicircle tiles between the double cornices linked with the horizontal lines. Projecting windows emphasize the horizontal composition dotted the division of the facade, and stress the loftiness of the apartments.

103

诺曼底公寓（今武康大楼，淮海中路1850号）由法商万国储蓄会投资兴建，占地面积1,580平方米，建筑面积9,275平方米，高八层，钢筋混凝土框架结构，是沪上最早的外廊式公寓。该建筑再现了法国文艺复兴时期的建筑风格。

Normandie Apartments

It was invested by a French firm, the International Savings Society. The eight-storey steel-concrete structure, covering 1,580 square meters with 9,275 square meters of building area, was the first veranda styled apartment in Shanghai. The building is a vivid representation of the neoclassical style of the French Renaissance period on a grand scale. Today as WuKang Building, is located at No. 1850 Middle Huaihai Road.

武康大楼旧称诺曼底公寓，位于淮海中路1836—1858号，一九二四年建造，钢筋混凝土结构，为上海最早的外廊式公寓，是邬达克的重要设计作品。一九九四年被列为上海市优秀历史建筑。张安朴作于丁酉春日四行储蓄会联合大楼（今联合大楼，四川中路261号）是邬达克独立开业初期的作品。虽仍使用新古典主义造型，但他不再局限于对西方古典主义建筑形式的模仿与重现，开始展现自己独有的特色和创造空间。转角塔楼据说是邬达克对匈牙利北部乡村式文艺复兴风格的追忆，他设计的大楼铜制卷帘门由布达佩斯的一个金属加工厂制作。

四行储蓄会联合大楼

Union Building of the Joint Savings Society

The building was built in the Georgian style, today as Union Building, is located at No. 261 Middle Sichuan Road. One of the early designs of Hudec's own firm, this building features neoclassical characteristics while at the same time reflecting the innovative skills of the architect.

郁达克设计的四行储蓄会联合大楼，位于四川中路261号。一九二八年建成，是上海首座纯主面大量采用大理石厚板的建筑。
张安朴 作于丙申冬日

上海啤酒厂

上海啤酒厂（今苏州河梦清园，宜昌路130号）原厂主营酿造业务，工艺流程全部机械化，所有设备从国外进口。厂区总平面做马蹄形布置，使土地面积得到充分利用。厂区主要建筑有酿造楼、灌装楼、仓库、办公楼和发电间等，曾一度是中国最大的啤酒生产企业。主要建筑为现代派风格。该厂是邬达克在上海设计的两个大型工业建筑之一，也是他将现代派新风格运用于厂房建造的成功实践。

Union Brewery Co.

The main business of the factory was a brewery. The process was controlled by mechanization and all equipment was imported from overseas. To make the full use of the land area, the master plan was designed in a horseshoe shape. The main structures included a brewery building, bottling building, garage, office building and electric-power room etc. The factory once boasted being the biggest beer producer in China. The main buildings were of Art Deco style. As one of the two huge industrial factories designed by Hudec, Union brewery Ltd. Successfully integrates Modernism with industrial design. Today as Mengqing Garden, is located at No. 130 Yichang Road.

上海啤酒厂住于宜昌路130号，是邬达克在上海设计的大型工业建筑之一也是他将现代派新风格运用于厂房建筑的成功实践。一九九年被列为上海市优秀历史建筑。

张安朴作于丁酉春日

大光明大戏院

大光明大戏院（今大光明电影院，南京西路216号）为典型的现代装饰艺术风格，凭借摩登和豪华，被誉为"远东第一影院"。它的建成，是邬达克在上海设计生涯中一个重要里程碑，实现了他设计风格的彻底转变。他对商业建筑设计的精准理解与定位、处理复杂地形与复杂功能的特殊能力、收放自如与现代时尚的表现手法，使其在建筑界备受瞩目。

Grand Theatre

The theatre is located at No. 216 West Nanjing Road, it has a modern Art-Deco style. For its extremely modern and luxurious characteristics, the Grand Theatre was reputed as "the best theatre in the Far East". As an important milestone in Hudec's architectural life in Shanghai, the completion of the Grand Theatre marked the turning point of Hudec's architecture style.

大光明电影院是邹达克设计的新潮作品之一，建于一九三三年，曾被誉为"远东第一影院"。
张安朴作，丁酉春日

达华公寓

达华公寓（今达华宾馆，延安西路918号）为现代派建筑风格，平面布局紧凑，立面造型简洁。该公寓从功能需求出发，摈弃烦琐装饰，利用简洁的体块和建材本身的色彩、质地及精美的加工工艺，表现出现代派建筑的形式美。

Hubertus Court

The Hubertus Court today as Da Hua Guest house, is located at No. 918 West Yan'an Road. It was designed in modern style with compact layout and simple facade stuccoed with milky cement. As one of the important projects designed in the last period of Hudec's architectural practice, Hubertus Court illustrated the beauty of modern architecture by emphasizing the functional requirements, discarding fastidious decoration, and making full use of simple volumes, colour and texture of materials.

达华宾馆旧名达华公寓，位于延安西路918号。邬达克于一九三五年设计、一九三七年建造。达华公寓为现代派建筑风格，平面布局紧凑，立面造型简洁，表现出现代派建筑的形式美。一九九九年被列为上海市优秀历史建筑。

丁国泰日

嘉定

南翔

苏州河

黄浦江

豫园

叁

·

豫园·嘉定·南翔

豫园湖心亭和九曲桥

豫园位于上海市老城厢的东北部,是江南著名古典园林,始建于明代嘉靖、万历年间。九曲桥九曲十八弯,可以通达湖心亭茶楼。

Yu Garden Lake Pavilion and Zigzag Bridge

Yu Garden, located in the northeastern part of the Shanghai old town, is an outstanding example of the famous classical gardens that were built south of the Yangtze River during the Jiajing and Wanli eras of the Ming Dynasty. The Zigzag Bridge in Yu Garden, with its nine turns and eighteen angles, leads visitors to the Huxinting Teahouse.

豫园湖心亭九曲桥

张安朴作于癸巳夏日

117

豫园大假山

豫园大假山由明代著名叠山家张南阳精心设计堆砌，是他仅存于世的作品，山高约14米，登顶可远眺黄浦江。

Grand Rockery

Zhang Nanyang, a famous stone garden artist during the Ming Dynasty, carefully designed this rockery, which stands as his only surviving work. At about 14 meters high, it overlooks the Huangpu River.

豫园大假山
张安朴作于癸巳夏日

豫园卷雨楼

卷雨楼与大假山隔池相望，后有回廊，曲槛临池，可观雨、赏花、戏鱼。

Juanyu Chamber

Facing the big rockery across the pond there is an ambulatory, a curved sill facing the pond that offers a perfect view of the rain on wet days, flowers, and fish.

豫园卷雨楼
张安朴 作于癸巳夏月

玉玲珑

玉玲珑为豫园镇园之宝，与苏州冠云峰、杭州绉云峰，并称江南三大名峰，具有太湖石的皱、漏、瘦、透之美。

Exquisite Jade Rock

This rock is the treasure of Yu Garden. With the furrows and fissures and taut, penetrating beauty of Lake Tai stone, it is the equal of the Guan Yun Peak of Suzhou and the Zou Yun Peak of Hangzhou, the three of which make up the famous Three Peaks in south of the Yangtze River.

玉玲珑
上海豫园是著名的
江南古典园林，全国
重点文物保护单位，
镇园之宝"玉玲珑"
具有瘦、透、皱、漏
之美，为江南名石，相
传是宋代遗物，
张岩朴作于
癸卯春日
沪西丰艺斋

豫园积玉水廊

积玉水廊南连会景楼，北达涵碧楼，因廊旁一石"积玉峰"而得名。

Jiyu Water Gallery

Jiyu Water Gallery connects Huijing Hall in the South and Hanbi Tower in the north. It is named after the stone "Jiyu Peak" next to the gallery.

豫园积水廊安朴写生笑已秋日

豫园门楼

砖雕的豫园门楼后面有江泽民题词的"海上名园"巨石。

Entrance Gate of Yu Garden

Inside the entrance gate tower of Yu Garden sits a large yellow-brown stone upon which Jiang Zemin inscribed the words "Famous Garden of Shanghai".

「海上名園」
專砖雕刻的
豫園門樓
張安樸作于
乙亥夏日

豫园
湖心亭

人们可从豫园门楼前的荷花池眺望湖心亭和整个豫园商城古群建筑。

Yu Garden Lake Pavilion

From the Lotus Pond in front of the entrance gate to Yu Garden, the lake pavilion and ancient buildings of the Yu Garden shopping market are overlooked.

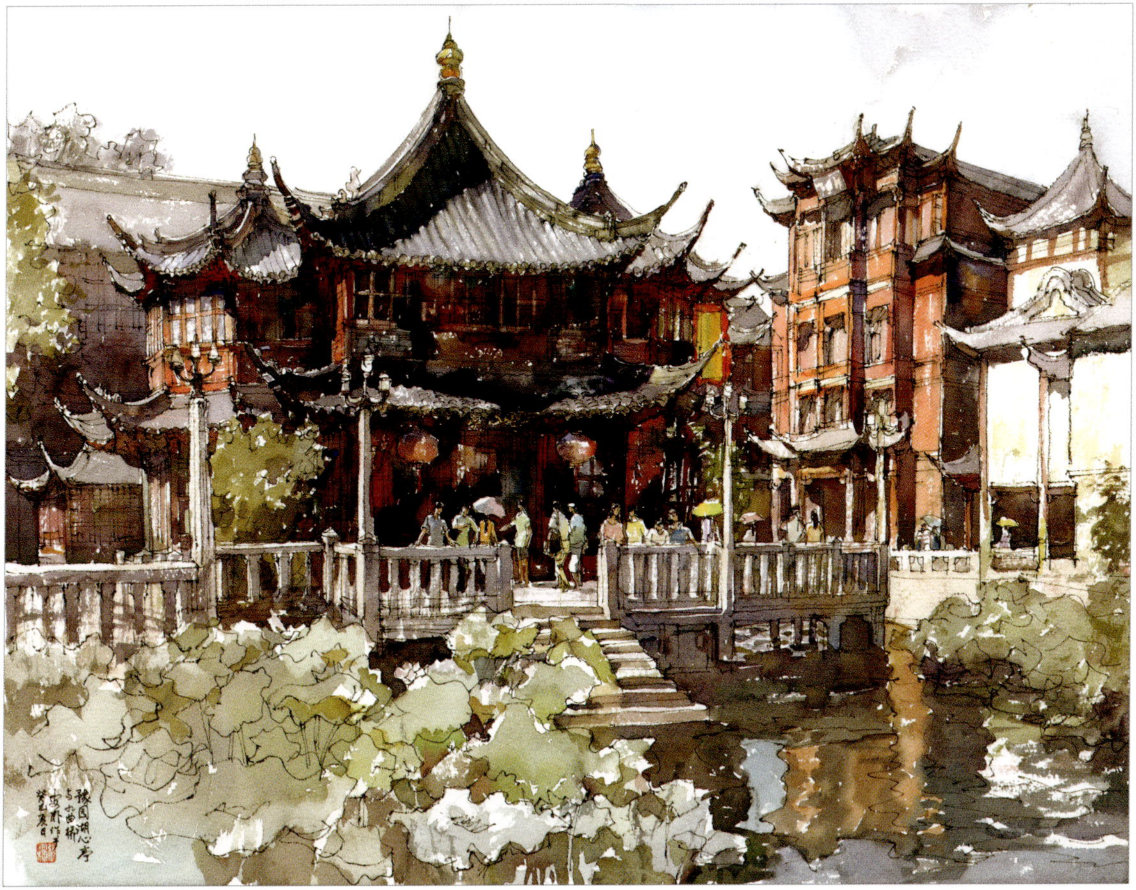

豫园湖心亭
与九曲桥
癸巳夏日

嘉定孔庙

嘉定孔庙位于嘉定南大街,始建于南宋嘉定十二年(1219年),大堂摆放孔子塑像和祭祀圣人时的器具,墙上绘有孔子与众学生的故事。

Jiading Confucius Temple

Located on South Street of Jiading District, the Temple was first constructed in the 12th year of the Jiading Era of the Southern Song Dynasty (1219 AD). Statues of Confucius and sacrificial vessels decorate the lobby, and scenes from the life of Confucius and his students are painted on the wall.

嘉定孔庙
位于嘉定城内,建于南宋
嘉定十二年,号称"吴中第七"
被列为全国重点文物保护单位
张安朴作于丁酉春日

汇龙潭

明万历年间，知县熊密在孔庙前五条河流交汇处开凿一潭，储文气，增景色，汇龙潭因此而得名。

Huilongtan

During the Wanli era in the Ming Dynasty, the county magistrate Xiong Mi dredged a pond in front of the Confucius Temple to preserve the culture and improve the scenery. Because five rivers converge here, the pond was given the name Huilongtan (five rivers like five dragons).

汇龙潭公园

明万历年间，知县健密在孔庙前开浚一潭，意为储文气、增景。龟将五条河流交汇于此，故命名为"汇龙潭"。一九八二年建成汇龙潭公园。

张安朴作于丁酉春日

秋霞圃是上海五大园林之一，以布局精致、环境幽雅、小巧玲珑的特点受到广大游客喜爱。

Qiuxia Garden

Qiuxia Garden, one of the five classical gardens in Shanghai, is loved by tourists for its exquisite layout, elegant environment, and daintiness.

秋霞圃，位于嘉定城内建于明代，为上海地区最早的古典园林。布局紧凑，富有诗情画意。现为上海市文物保护单位。张安朴作于丁酉春日州桥老街位于嘉定西门老街一带，河道纵横，水系发达，保留着明清古民居特色。

Zhouqiao Old Street

The Ximen old street area of Jiading, with its intricate waterway network of canals and small bridges retains the characteristics of the traditional homes during the Ming and Qing Dynasties.

卅橋臨河的老街老屋
古樸作
壬辰秋日

法华塔

法华塔位于嘉定区州桥老街，始建于宋代开禧年间，为四面七级楼阁式砖木结构方塔，登顶可眺望全城景色，是嘉定城区标志性建筑之一。

Fahua Pagoda

Fahua Pagoda is located in Zhouqiao Old Street in Jiading District. One of the landmark buildings in Jiading, it was initially built during the Kaixi period of the Song Dynasty. It is a seven-storey brick-wood square structure with surrounded by pavilions. From its top, you can overlook the scenery of the entire city.

嘉定法华塔

法华塔又名"金沙塔"、"文笔塔"，为嘉定标志性建筑。始建于南宋，七级方形楼阁式建筑。登顶可眺全城景色。历代多次重修法华塔。一九四年又全面修建，现为上海市文物保护单位。

张安朴作于丁酉春日

品小笼 长桌街宴

五湖四海来相聚，千桌万人品小笼。每年9月，上海南翔小笼文化展盛大开幕，千桌万人齐聚南翔老街品小笼。

Shanghai Dumplings served on the old street

Every September, the cultural exhibition of Shanghai Dumplings is held in Nanxiang. People from all over the world gather in the old street and taste the delicious Shanghai Dumplings.

141

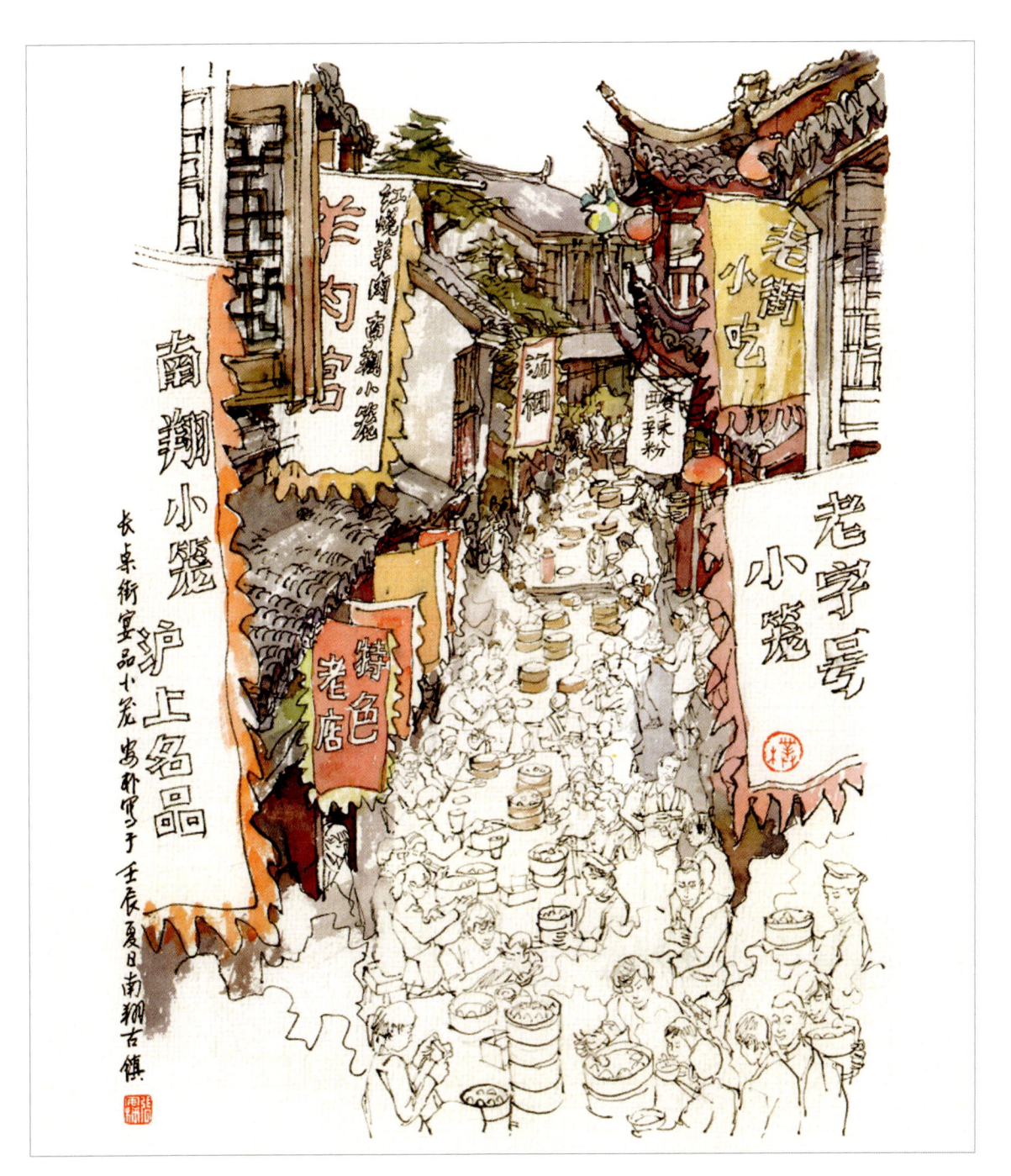

南翔老街人气旺

千年古镇万人聚,喜迎八方游客来。南翔老街是南翔古镇的精华所在,向游客娓娓讲述"银南翔"的商贸、文化和历史。

The flourishing old street of Nanxiang

Thanks to its rich commerce, culture and history, the old street is always the center of the ancient town.

南翔老街人气旺 安朴作于壬辰夏月

南翔双塔

千人聚双塔，吟诗赏烟霞。南翔双塔建于五代至北宋年间，是我国古砖塔中的罕见珍品。火焰形的壶门、简朴的直棂窗、精巧的斗拱、挺秀的塔刹，向世人展现南翔双塔建筑工艺之精美。

Nanxiang Twin Towers

Built between The Five Dynasties and North Song Dynasty periods, the twin towers have become rare treasure of ancient Chinese towers with the flame-shaped decorations, the vertical bar windows, the elegant brackets and the graceful tower spire demonstrating the craftsmanship of Chinese architects.

南翔双塔

南翔古镇
八字桥

八字桥指的是吉利桥、太平桥和隆兴桥，三座桥横跨横沥河和走马塘两条镇上的主要河道，呈八字形挨在一起。南翔人至今还有走三桥的祈福习惯。端午竞渡思先贤，走过三桥百病除。

The Bazi Bridge

Bazi Bridge, ("Bazi" refers to the Chinese character "八"), which is made up of three bridges, including Gili Bridge, Taiping Bridge and Longxing Bridge. According to the tradition, one must go across the Bazi bridges during Dragon-boat festival. After crossing all the three bridges, he/she can be relieved from all sicknesses and bring home good luck.

南翔古镇"八字桥",由三座桥构成一个"八字"俗称"八字桥"。张宝林作于丁酉春日

南翔天恩桥

烟气笼南翔，流文荡天恩。天恩桥是上海市十大古桥之一，是现代立交桥的雏形。

The Tian'en Bridge

The Tian'en Bridge, one of the top ten ancient bridges in Shanghai, is the archetype of modern overpasses.

南翔天恩桥

住于南翔镇永丰村，始建于清顺治年间，是嘉定现存最大三孔石拱桥，桥长50米，横跨横沥河。"天恩赏月"被列为"槎溪十八景"。

丁酉春日 张安朴作于

巴基斯坦馆、尼泊尔馆、以色列馆、阿曼馆、阿联酋馆、新西兰馆、马来西亚馆、新加坡馆、澳大利亚馆、印度尼西亚馆、泰国馆、国际组织联合馆、太平洋联合馆、红十字与红新月馆、世界气象馆、联合国馆、国际信息发展网馆、亚洲年馆、比利时-欧盟馆、瑞士馆、波兰馆、法国馆、德国馆、英国馆、荷兰馆、意大利馆、卢森堡馆、奥地利馆、芬兰馆、斯洛伐克馆、捷克馆、丹麦馆、瑞典馆、挪威馆、欧洲联合馆、船艺大厅城市博物馆、韩国企业联合馆、城市足迹馆、日本产业馆、石油馆、国家电网馆等。

宋扑写于二〇一〇年八月上海世博会园区

二〇一〇年上海世博会园区面积为5.28平方公里。展馆包括主题展、国家馆、国际组织馆、企业馆等。此鸟瞰图有世博轴、世博文化中心、大陆各省市馆、全景、香港馆、澳门馆、台湾、公共展厅。中国国家馆、世博中心、世博主题馆、

肆

世博园区

中国国家馆 上海世博会

中国国家馆以城市发展中的中华智慧为主题，表现了"东方之冠，鼎盛中华，天下粮仓，富庶百姓"的中国文化精神与气质。世博会结束后，该馆更名为"中华艺术宫"。

China National Pavilion of Shanghai World Expo

Taking "Chinese Wisdom in Urban Development" as its theme, the China Pavilion displays the Chinese cultural spirit and temperament captured by the expression "the Crown of the East, the flourishing of China". After the World Expo, it was renamed China Art Museum.

上海世博会中国国家馆于二〇一〇年春天全面建成。中国馆外观以东方之冠、鼎盛中华、天下粮仓、富焦百姓为构思主题，表达中国文化的精神与气质。二〇一〇年五月一日上海世博会隆重开园，迎来四方宾客。安朴记于庚寅初夏

153

上海世博文化中心

上海世博文化中心造型呈飞碟形状，白天如"时空飞梭"，似"含珠巨蚌"，夜晚则梦幻迷离，如"浮游都市"。现更名为"梅赛德斯-奔驰文化中心"。

Shanghai World Expo Cultural Center

Shaped like a flying saucer, this magnificent structure looks like a giant clam with a pearl during the day and is dreamy and blurred at night. It is now called the Mercedes-Benz Arena.

世博文化中心造型呈飞碟状，白天犹如"时空飞梭"，似"合浦巨蚌"；夜晚恍偈梦幻迷离，恍如"浮游都市"。世博文化中心集演艺展示娱乐于一体，内部可转换成各种舞台和场所，供各类活动需写。安朴写于庚寅盛夏上海世博会园区

上海世博会 俄罗斯馆

展馆由12座塔楼组成,顶部为镂空的俄罗斯民族服饰图案,如12位美丽的俄罗斯姑娘在跳舞。

Russia Pavilion of Shanghai World Expo

The Russia Pavilion is composed of 12 towers, the upper parts of which contain perforated patterns inspired by ethnic Russian costumes giving the impression of 12 beautiful Russian dancing girls.

俄罗斯馆由12个塔楼组成。每座塔楼的花饰来自各民族扣女的传统服饰图案。整个设计就像12个姑娘在跳舞，充满着生命的象征而别具风彩。

学朴写于庚寅夏日

上海世博会
卢森堡馆

展馆的建筑结构像一座堡垒,把一座塔楼包围其中,周围是郁郁葱葱的开放式树林。

Luxembourg Pavilion of Shanghai World Expo

The structure of the Luxembourg Pavilion is like a fortress in a forest, built from welded rust red steel plates and surrounded by lush public-space trees.

亦小亦美的卢森堡饭，似用巨石雕凿出结构如壁垒般的建筑，周围则是郁郁葱葱的葡萄园。

安秋平于庚寅春日

图书在版编目（CIP）数据

手绘上海经典建筑 : 汉、英 / 张安朴著. -- 上海 : 上海人民美术出版社, 2024. 12. -- ISBN 978-7-5586-3058-3

Ⅰ. TU-862

中国国家版本馆CIP数据核字第20242ML969号

手绘上海经典建筑

著　　者：张安朴

责任编辑：朱卫锋　张　璎

技术编辑：王　泓

版式设计：钮清越

制　　作：顾　静

文　　字：王兆煜

出版发行：上海人民美术出版社
　　　　　（上海市闵行区号景路159弄A座7F）

印　　刷：上海丽佳制版印刷有限公司

开　　本：787×1092　1/20　8印张

版　　次：2024年12月第1版

印　　次：2024年12月第1次

书　　号：ISBN 987-7-5586-3058-3

定　　价：98.00元